Other b

Empty Thoughts frc

Spaced Out and Cut Up

Stupid Jokes for Clever People & Clever Jokes for Stupid
People

Observations from Another Planet

~~~~

Books in the Logic List English Series

Logic List English – Rhyming Words etc. - Volume 1A
Logic List English – Spelling Arrangement – Volume 1B

Logic lists English - Multi-Syllable Words - Volume 2A
Logic lists English - Split Multi-Syllables - Volume 2B

More Volumes Coming

THAT'S THE STATE WE'RE IN!

A collection of Short Stories

BY TONY SANDY

DragonEye Publishing

THAT'S THE STATE WE'RE IN!
A collection of Short Stories
Copyrighted © 2017 by Tony Sandy

First Edition
First Printing May 5, 2017

ISBN 13: 978-1-61500-146-0 (Paperback)
ISBN 13: 978-1-61500-102-6 (ePub ebook)
ISBN 13: 978-1-61500-182-8 (PDF)

Library of Congress Control Number: 2017940167

Published by Social Misfits, an Imprint of DragonEye Publishing

Contact info.
www.DragonEyePublishers.com
Orders@DragonEyePublishers.com

DragonEye Publishing
753 Linden Place, Unit A
Elmira, NY 14901 USA

Introduction

This book is a collection of short stories, based upon various human conditions or states. Some are just sketches, detailing what it is like to have these medical and psychological problems in your life - others are more involving. Enjoy or not as the case may be but I hope you have your thought buds at least awakened by the ideas within.

Contents

THAT'S THE STATE WE'RE IN!
A collection of Short Stories

THE LAMENTATION OF JOB*

"Why have you done this to me? Why? Why? Why?"
After a few agonised seconds, the voice went on.

"Have I not served you loyally, all these years? Have I not done everything you commanded of me!?! Yet you've destroyed my family, my world and now my own flesh!

What more do you want of me!?!" The voice demanded angrily.

"You've made my life a misery and still expect 'gratitude' from me because I'm not dead yet! What kind of monster are you? Even The Devil himself wouldn't torture me this way. Lord of all things - including my pain and hatred.

Oh yes I loathe you as once I loved you. You were my God. You could do no wrong in my eyes but what right have I done in yours? You've destroyed me. You've crippled me with physical pain and emotional misery. How can any sane man do anything but hate you with every fibre of their being? How can I do anything but treat you with the contempt you've obviously poured on me?"

Suddenly a voice of great majesty and mystery boomed out.

"Oh you foolish mortal! How little you know of me and my works.

Who are you to question me and what I do? What knowledge do you have of how Creation works. Who are you really Job? Yet you stand against me, bleating like a little lost lamb. Does this not condemn you as the child you are? The world is rebuilt every day by faith alone. You foolish man - you have become lost in the illusions of this world. It is not the built that matters but the building of reality at every instant of existence.

The man that holds onto wealth and power kills it, crushes it. Like sand running through his fingers, it is gone in a moment. Fear of losing it, washes it from his grasp. Woe to you, you doubter! Your sin is in your heart that you do not believe. You are lost in false pride! Where is the humility of the servant? Where is the joy in serving the greater good, through your fellow men? Lost in selfish self-aggrandisement!

Why do you no longer search for me in the wild places? Am I not God of all things, after all? Do not the trees, the plants, the animals - the very ground itself, praise me with its very existence? Yet you despise me and turn away from my goodness, my bounty for you."

Suddenly a new voice burst forth, spitting out its deep contempt.

"You would drown your sorrows in the tavern, if you could lower yourself to mix with the other offal of society. You are tempted but won't betray this illusion of false pride, even now. You visit the Temple as a wealthy merchant - no man must accuse you of the vain hypocrite you are, oh no! You think we do not know who you really are, under this festering facade of self-righteousness? You think we do not see how black your heart really is? You spurn those you should be helping, instead falling into my hands because of who and what you are. I am God's prosecutor and you are guilty of error before him.

Have you ever turned any other man from sin, by pouring out your genuine heart to him?"

"I have given to charity."

"You have thrown worthless coinage into the gutter. Have you ever given yourself, to the hurt, the emotionally lame? Have you given hope to but one, single man?"

Job looked down at the ground shamefacedly.

"No, I thought not. This is your guilt. You've watched your fellows rolling in agony on the ground and walked contemptuously by. The prosecution rests, Lord."

Another, more gentle voice now spoke forth.

"We are not here to punish you for your sins but to alert you to them. God isn't punishing you. We don't have to punish you for your sins - you are already doing that yourself. God didn't take your family - you let them go. You failed to protect your children that is all.'

A great wail went up to the heavens. Job realised his error and his pain.

"Oh Job, our heart goes out to you.

You are more than you think you are. It is you who has made yourself ill, with this self-doubt. Your inner turmoil is reflected in your outer symptoms. How can you claim to know God, when you don't even know yourself?

We are not here to judge you - you judge yourself. You find yourself guilty or innocent and execute the sentence upon your own body or mind. You are your own judge, jury, executioner or jailer.

Why do the righteous suffer? Why do sinners get away with murder? Because they sin by their acts and blame the outside world for the failures. Their sin, their error is internal - in their own thoughts. The righteous suffer in silence, not because they want to but because they must. The world is made by the tolerant and patient, and destroyed by those with neither of these qualities. God doesn't punish - how can he, for he is love personified?"

Another great cry of pain went up from Job as he understand what he hadn't before.

"It is our foolishness that punishes us. In harming others, we harm ourselves. They are children in relation to the adults that rebuild this world anew each day, blessing it with their joy. Every disaster throws down this ants nest of a world and the long suffering rebuild it because they must. Those who give into the pain and misery, give up this world and drift from the path of righteousness, like leaves in the wind. They become children again - alone and abandoned by God, in their eyes own eyes at least. It is they though who have shunned the light - wandering from the straight and narrow way.

The Lord is a mirror. Turn your back on him and he'll turn his back on you - not because he wants to or does in reality but because life is voluntary. You are here because you want to be and are free to leave at any time. Give up all effort and you will drift away from life naturally.

Fear the law, not the Lord.

Man interferes because he does not understand God's laws. When he does understand, he steps back and lets those laws work automatically. Sinners are new here. One day they will realise that this world is theirs. At that point they will stop sabotaging their own good fortune - killing themselves in the tortured belief that they are killing an enemy."

At this point Archangel Michael fell silent. For Job this manifested as a high pitched whistle in his head, reflecting the emptiness he now felt that his struggle was over. In time his body healed itself, his wife forgave him and bore him a new family. He forgave his friends their foolishness as he forgave himself for his own. His life prospered again in all ways because he had returned to the fold and was his true self once more...

(*Inspired by The Sire of Sorrow by Joni Mitchell)

THE MAN WHO KNOWS EVERYTHING

So this is Wednesday or is it? Maybe it's Thursday and by some bizarre coincidence, they've put the clocks back by a day? Is it a Leap Year? What is Wednesday anyway but an artificial construct, created by somebody as crazy as me? Maybe they're not crazy though? Am I crazy? I could be I suppose but how do I know? How do I find out for sure? What is the criteria for measuring insanity? Is talking to yourself a sign of madness? Am I talking to myself now, just because nobody is visible in my vicinity? Do voices in your head reflect imagination or telepathy? Do I know the answer to anything for certain? Does anyone?

I used to think the sky was pink but somebody told me it was yellow? Do I believe them? How do I define pink? What do they mean by yellow? It's all comparative isn't it?... Or is it? What if that door over there is yellow to me but red to someone else? What if all the street doors seem to be all the same colour to me but other people see them as multi-coloured variations on a theme? I suppose I could ask them how they perceive the colours but they could lie, couldn't they? And if I asked them directly, if they saw the same thing as me, they could say yes but it may mean something totally different to them, than it does to me. Apparently the Japanese have 83 different meanings for the word yes and none of them mean 'yes' in English. Of course I could be wrong and it might be 43 or even 42 (Where have I heard that number before and in what connection, one wonders? Maybe I haven't heard it before, just imagined I did? Isn't memory wonderful? I sometimes wish I had one!).

6

THAT'S THE STATE WE'RE IN!
A collection of Short Stories

Yesterday upon the stair
I saw a man who wasn't there
- me I suppose!'

I'm a 'Nowhere Man' but then again, maybe I'm not - just I think I am. Who can tell? Certainly not me.

My wife used to say I was a strange one...or did she? Was I ever married? I don't think so. I have no memory of it - only of having lived alone, on some vast plain, with nobody else around for miles, if at all but maybe I'm wrong on that. I seem to be surrounded by people and houses now but am I? It certainly looks like it but how can I be sure it's not an illusion? Maybe life really is like 'TheTruman Show' or 'The Prisoner' and I'm caught up in some vast conspiracy, aimed at confusing me. Then again maybe I'm deluding myself and it's not. How can you tell? Can you tell? Am I just a figment of my own imagination or somebody else's? One day I may wake up and find out I don't exist. Still that's much like any other day for me, isn't it?

Nothing bores me, more than me. I question everything because the universe is more fascinating than I am. Prancing around in front of others, pontificating as though every speck of dust on my nose hairs, meant something really, really important - that kind of thing gets really up my nose, in no small respect, if you know what I mean. I used to be like that. I was a prince or something like that...or maybe I wasn't. I've got an active imagination, with no foothold on reality, my teacher used to say. I'm pretty sure I had it once but it was the drugs that really blew my mind or was it enlightenment in that ashram? Who knows, it was all so long ago or so it seems. Maybe it was yesterday. Who can tell? Where does the past leave off and the present begin? Am I really here, wherever here is? Is here, here - does it exist or is it a mental projection of a place I once knew or wanted to know? Is it imagination or memory? Who am I to ask anyway? In fact who am I? I seem to have asked that question recently. Did I though? Does that mean I have a memory or is it that my imagination is misleading me again? Oh God, I never seem to resolve anything or perhaps I do and don't realise it? How could I tell? I need to keep taking the pills. What are pills anyway? Do they work because of what they are physically made of or because of the effect you're told to expect and then believe in, so thoroughly that your mind makes them work? Do I really know the answer to this? No. Do I have any pills? No.

It looks like it is windy outside. Is it? Am I outside? Looking at the evidence and considering the premise, I'm not sure I am. Am I in the premises or are the premises in me? I don't know and could be wrong about anything and everything. My university professor once said that you can't be sure of anything but whether he meant 'me' specifically or anyone in general, I never did find out as shortly after he was dragged off to the local mental asylum, claiming 'they' were out to get him and you couldn't trust anyone as 'they' are all over the place. The proctor of the college I was in, said I was all over the place too and sent me down shortly after. Funny old world.

My Head Teacher told me I'd never make anything of myself and the woodwork teacher said much the same thing

'You haven't got the right tools to make anything - in fact you're the biggest tool in my workshop!' At this everyone laughed, including me. He grimaced at me. 'You're not supposed to join in but feel ashamed at your crassness!' he blustered.

'Get your head out of the clouds - unless you want to be an astronomer or a pilot!' Was the sage advice of my science tutor.

'Stop dancing on cloud nine' the music teacher said to me.

'Wallop!' The gym instructor's shoe advised my ear.

Or is this all imagination? Was I ever really a child? Maybe I was cloned and all these 'memories' are hereditary, from the selves I used to be, will be or could be? It is all too confusing, to be sure of anything. There is just too much going on in my head and not enough in my cranium, to ground me.

A great philosopher once told me that 'must be' thinking leads everybody astray. It is believing you have the answer to a problem and that the solution 'should' work but doesn't. Your mind insists you keep trying, until somebody else with the real answer comes along and switches the light on because they have no belief about how things should be, only how they are. Einstein said much the same thing as did Mrs Baker at number 32, when I was a kid...or did she?

THAT'S THE STATE WE'RE IN!
A collection of Short Stories

Life is so shallow, so boring at times and this seems to be one of them. I sometimes think that I've found the answer to life, the universe, everything - then it simply slips through my fingers (Funny, that number 42 keeps popping into my mind). Maybe we live in a multiverse and all these characters bouncing around in my head, are me too but based somewhere, some-when else? Who knows? Who cares? Not me today, that's for sure. So life is 'fascinating,' interesting, full of facts? God how I hate them sometimes - 'facts' that is! Maybe that's why doubt creeps into my mind all the time, just to relieve the tedium of certainty - that blank canvas, which is full of things but mentally a desert (No stimulus, no challenge - nothing to take you away from this dusty museum of relegated relics of the mind, carefully stacked, categorised and forgotten).

When I talk to myself, who is this me I speak to? Is it some vast audience of passive receivers, out in the depths of space, so bored out of their minds that they've got nothing better to do with their time, than listen to my ramblings? Maybe it's the cells in my own body, tuning in to the dictates of their giant leader and being sadly disillusioned by their sheer inanity?

'Conscience doth make cowards of us all,' Bert the gardener told me once, which I thought was a bit deep for him. I've noticed that people with big egos, tend to tread on the toes of other people with little egos. They are just so full of themselves that they don'y notice anything but their own reflections. What I believe pain is, is these little voices screaming in unison. I've been walking along, full of myself, when I accidentally step on a snail that was also full of itself, until this happened (Like a hedgehog hit by a lorry, it's now empty of itself - squashed, flattened, insides outside): Crushed shell, crushed ego. Poor snaily-waily! Still, what do I know about fate and the destiny of the universe, in relation to this little fellow's demise? Chaos theory rules or it would, if it could hold a pencil straight.

Hmm. What is that sort of emptiness I feel? That sort of rumbling below the belt of my trousers, calling my attention? What does it mean? What is it trying to tell me? Is that vast, aching, cavernous vacuum, hunger? Should I give in to it and eat something or resist the urge? Is it right that this organ, which is only part of me, should dictate to the whole of this thing I call 'me?'

Oh dear, now what is this? It sort of burns, like acid. Is it acid? If so, what is it doing inside of me? Should I call a doctor or put my fingers down my throat and try to bring it up? No. I remember being cautioned about doing that, if you've drunk battery acid (People on this planet seem to swallow anything - is that why they have a car boot sale, I wonder?). Drink milk - yes that was it! Where do I get milk from? A cow. Where do I find a cow at this time of day? My mind seems to be telling me to open this big, white, humming box. Is it a Tardis? Is there a whole cow inside and if so, do I know how to milk it? Oh dear, all these questions! I need to be decisive. Hello, is anybody in there? If there is I'd like to borrow some milk please - you know that yummy white stuff that comes out of those black and white things with horns. Hello? 'Bang, bang!' (He hits the door and it swings open).

THAT'S THE STATE WE'RE IN!
A collection of Short Stories

Oh, it's just a fridge, full of cold things, like milk I hope. Are yes, that will do. Pour this white stuff from its plastic container, into this clear container, which I'll call a glass. Oh look, it's turning white too! Must be the milk. I'll raise it towards my head and stick it in this thing that keeps opening and closing, which I'll call my mouth. Now that's better. The burning sensation has stopped but I need something more solid, to stop this rumbling, my stomach is telling me. Should I give into it? My next door neighbour says I shouldn't let anyone else push me around. Is she right? Should I listen to her? The kindly old lady from number 42 says I should let my heart be my guide. Does that mean I shouldn't let my stomach bully me into submission? It's all so complicated - who should I believe? What should I believe? What is the actual truth and what is solely supposition? I wish I knew. One man's truth seems to be another's lie.

What's that smell? Is it burning? Must be the toast! Did I put some on? I suppose I must have as I appear be the only person in this pile of bricks, people call a house.

Anyway I'd better butter this bread and plunge it into that deep, dark chasm, beginning at my mouth. Umm - what is that kind of hard but crumbly texture? Must be toast. And all those funny little bits, cascading off it - must be sand. No, can't be that. I remember getting sand in my mouth, during a trip to the seaside. No, it can't be that - this isn't as tooth-breakingly hard. Crumbs, what is it? Ah, that is it - crumbs!

I remember my first experience of sex as I think they call it (In fact my only experience of it). The girl thought it was wonderful because I went on for hours without 'coming,' I think was the word she used. It was the same when it came to using toilets and not being able to urinate in public (When do I breath in? When do I breath out? What's the trigger point?). It was like all my attention was being drawn outwards because I didn't feel safe enough to relax and let go, unlike in my own environment. Doctor Woods called it performance anxiety, like when in the school play and drying. My school compatriots said it was just because I was a weirdo. Talking about all this makes me feel anxious, like when I had all these strange experiences they call psychic, I believe. Maybe I am a weirdo, a freak that just doesn't fit in with society. My mother (I suppose I must have had one) said I was unique, not a freak (Unless I came out of a test tube). My father replied 'I dunno, lot of flying saucers about nowadays.'

Did I open that door earlier or did it open itself, I'd ask myself as a child. I remember once finding myself downstairs at night and the door into the living room just creaking open. I screamed the house down apparently. My mother said I'd been sleep walking. My dad said it was the cat that pushed the door open but we didn't have a cat. Then there were the times I caught movement out of the corner of my eye. This girl I knew, who was as equally strange as me, said I was seeing ghosts or things from other dimensions. Her mother called her gifted, being a spiritualist herself. Her father just said she'd been dropped on her head as a baby. Funny about fathers and their funny explanations about everything. I went to one of their meetings once (Spiritualists, not fathers). They told me I was very gifted too. I couldn't bear it as I couldn't tell if what I was thinking and feeling, were my thoughts and feelings or someone else's (All these words, ideas and ailments suddenly flooded my body. I'd just have to look at someone to know what they were thinking or tell from the discomfort in my own body, where the illness that plagued them, was based. I couldn't stand the noise in my head and the sensations in my body, so ran out. It's the same with pubs, gigs and discos. Noise inside or outside my head, was more than I could put up with, so being alone was the only thing that saved me or could save me. In an earlier age I might have become a monk, hermit or a shaman. In these enlightened times I've become a recluse. I became Doubting Thomas because Certain John couldn't stand the attention. I freaked out and backed out of life because it drove me crazy. Now I sit here, calm, cool, collected - questioning everything, including my right to exist and the belief that I do...or do I?

ARCHIE IN THE FIFTH DEMENTIA

Well here I am again, not knowing where the hell I am. Everywhere looks the same - every house, every street, every town. Only the past remains my home. Only it has any distinguishing features. This place, these people mean nothing to me. They don't really see me and I don't see them. Even these people who say they know me - I don't know them. They say they're friends, relatives, colleagues from work. One even says she's my daughter but I don't know her from Adam. I can't remember names and when I can it's like a fruit machine spinning its wheels. Is it Dave, Doris, Dick, Arthur, Edward or Arnold? Eventually the right name comes round or they blurt out the answer.

'It's Mary, your sister'

'Now, Mr Smith - what do you want for your tea?' a voice breaks into his reverie.

'Is she talking to me? Am I Mr Smith?

'There's apple crumble or rhubarb tart for dessert and shepherds pie for the main meal or your favourite - macaroni cheese'

Is it? What does it taste like, I ask myself?

It is sad when I look back on all I was capable of and now I can't even tie up my own shoes laces. I think of all the places I've been in the world and now I'm trapped in this single building.

Everyday is the same here. They call it a home but it feels more like a prison. The residents who've still got a mind, are not as badly off as the others who haven't. 'The Wanderers' are like wind up toys, changing direction when they bump into an obstacle. They seem to run on Duracell Batteries because they never stop, night or day. Others, mostly staff, call them less pleasant names, like 'zombies' or 'the living dead.' Still maybe that's their way of coping with the tiring motion of the totally ungrounded. I am lucid most of the time but these 'vegetables' as one particularly sadistic carer named them, never are. They don't talk, just make noises, if any sound at all. They are no different from those kids euphemistically dubbed as having learning difficulties. They scream, they shout, they have temper tantrums and call for their mothers, long since dead (They are now their own parents, in body at least - if still children at heart). Again, you can hear the same cries of despair within the walls of a prison but from true youngsters, waking up to the hell that their lives have become.

Looking out the window, I see mashed potato clouds floating by. Yesterday they were like sailing ship, their bottoms gouged out on yonder mountain tops, wrecked on the reefs of open despair. Even if I cannot escape this prison physically because my body is too weak, at least my mind can fly free on wings of imagination.

We are all dreamers here. We dream that we are children playing in the woods and fields - swimming in the rivers of yesterday. Our future is your past and our past, your future. We are the husks of what was - the skeletons of what might have been. We are the dusty plains and you, the fertile jungle, running riot. We are the old river, silted up and running dry of energy - the rocky promise, now empty.

All you see, when you look at us is a vacant tent - worn canvas, stretched over a rickety frame. The wind whistles between our years, making sounds few can decipher or want to.

Outside, foam towers float down the dark river, like icebergs on a sea of trouble. Last winter it burst its banks, sweeping them clear like some giant broom, combing the reeds and weeds flat. 'How do you like your hair sir?' asked nature's hairdresser, washing away all protests with a wave of its hand.

On the land, an ocean of corn is awash with green straw that drowns my imagination.

I remember twisted bodies in my youth, writhing in agony - lovers or the dying? Oh blotched body, where is your unblemished skin of yesteryear? Here I am today but where exactly is that? Mary, where are you when I need you? Long gone as I soon hope to be.

Where are my old friends, my wartime buddies? Gone too, mostly - some well before their time,

either mercifully with a crack of a bullet or unmercifully torn to shreds by a landmine or falling shell. One minute a recognizable human being, the next a blood spattered pile of offal : A hand that once shook mine, a foot that walked beside me or worst of all, a detached head or an unrecognisable face, torn off in seconds, that an instant ago was a live human being, a name you knew. I threw up the first time. After that the shock paled my skin and stilled my heart and tongue. Now I can't wait until I join them all, once more in the past I knew and loved. This sterile world holds me in thrall still, though - until it lets me go at its whim, not mine.

We who are about to die, salute you as you replace us on the conveyor belt of life and we slide inexorably towards its edge, eventually to tip into the dark pit of unknowing. Our minds precede us and our bodies come slowly crumbling after. This is our fate and your inheritance. You are fresh troops in this war of survival. Welcome to the chaos and confusion that is the battle of life. Take your seats for the journey that never really ends - just seems to as we fight towards the light and drift back into the dark. Know me now as I lose sight of you and become an empty symbol. The rocket is burnt out. The light and life's gone - vacant of me and those I loved and lost...

IT'S A CRISPY, CRUNCHY SORT OF DAY

God, that wind is bitter, in a yellowy-pinky sort of way.
need to put something warm and woolly on. Woolly - hmm...
that has a nice, kind of grey sound to it.'

Mike walked out the door as a gust of cold wind hit him
in the face.

'That whistling wind was in A flat - you'd have expected
it to be in C sharp!' he laughed to himself.

'Oh well, this won't get the cow milked or the beer
barrels down the cellar' he said to himself and walked on.

'This is definitely hot chocolate weather' he thought as
he waltzed on down the road.

'Morning, Mrs Wilson!' he grinned at the approaching
woman.

'Morning Michael! Cold day!' she replied.

'Yes but there's a tinge of orange in the air'

'If you say so'. A look of 'Oh my God, here we go
again, crept across her embarrassed face, which turned
slightly away at this, too her, unfathomable remark.

Michael continued 'You can almost taste it, when the
wind blasts you in the face'.

'Can't argue with that' Mrs Wilson said in a bemused,
rather than amused way. She looked at him strangely,
knowing this was just part of him and his strange family. His
father for instance and the way he referred to all lawnmowers
as being 'Grahams' - moaning, tough, stubborn and always
arguing with you, rather than just cutting the grass as they
were simply meant to do. 'Nutters, all nutters' she thought to

herself ' -but mostly harmless'.

Michael moved off. 'Well, shopping to do, must dash! Goodbye Mrs Wilson'.

'Yes' she said, then realising she was looking at him strangely, she shook herself free of the self-induced trance and said 'I must go too'. Turning her back on him, she walked off as well. 'Why is it always like talking to the Wooden Tops, when I meet that boy?' she muttered to herself with regret, before disappearing down the street.

'Must remember to get some 92's in Tescoes' Michael thought to himself. 'I really love that flash of red, which the number 92 brings to mind (92 evoked scarlet in his thoughts, which was a code for peanuts, for no sensible reason he'd ever been able to discern but it was always 92-scarlet-peanuts as an association in his head).

Great uncle Ernie was funny. He used to say that Tuesday made him sad - not the day itself but the name. All he had to hear was the name repeated, to feel himself slipping into a bleary eyed state. Kids at school would take advantage of this to make him burst into tears by chanting 'Tuesday! Tuesday! Tuesday!'

'What's up Smith?' a teacher would innocently ask.

'They've been screaming Tuesday at me again!'

To which the teacher would respond by clipping him round the ear, saying 'Pull yourself together boy!' or a baffled 'What the hell are you on about?' again followed by a clip round the ear, for being cheeky to the teacher. 'I don't know what your game is but I'm not putting up with this utter dribble - do you understand? See me after school.'

'Yes but-'

'Don't but me or you'll have to stay behind for longer!'

When he was older, he used to say a bit of how's your father, smelled of chestnuts at the crucial moment, for some reason ('Could be to do with Christmas and stuffing!' he always said with a wink).

As a small child, he apparently nearly jumped out of his seat at the table because an orange's bitter taste, made him see a bright, green flash in front of his eyes.

'I've always been wired up strange' he always said, when reminded of this and other incidents, that elicited peculiar responses in him.

Uncle Dick was totally different. He became a maths teacher. He said that he was so good at it that numbers jumped up at him. Colours and geometric shapes, stood out like 3-D versions of the real 2-D things, floating in space before him, like what we'd call holograms I suppose. He also used to say that sound was a pyramid and that high pitched notes were at the pinnacle and low, bass sounds were naturally 'at the base.' He also said that the pyramid was coloured - the bottom being dark and physically heavy. As you went up it, it got lighter and lighter, going through all the colours of the rainbow (and then some), until you reached the very top which was bright and 'light,' in both senses of the word.

He also used to say that he could see everything in layers

or scales, like music (He even said emotion was the same); you name it - it was all graded by opposing forms of existence, from hard to soft, up to down, inside to out, fast to slow etc.

He further said that males were hard, rigid, spiky, crude and the feminine was marked out by being soft, round and refined. Women were fluid - melted by their own warmth, into one form or worn smooth by time and motion. Men on the other hand, were rigid crystals - separated by their cold,hard, incisive attitude to life. Women were settled - coagulated into a rounded form, like The Earth itself but men were always unsettled and unsettling (up in the air).

It's funny but he was a genius that way because I was useless at maths myself, plus couldn't tell right from left and was always getting lost as a child because I had no sense of direction.

Auntie Jane was a brilliant artist and the only member of the family to get anywhere, apart from Uncle Dick. She was a mean piano player too, preferring Blues and Jazz because they inspired her so much, with their dark, opulent colours, she said. 'Oh those rich, velvety, jazzy purples,' she'd croon, when tinkling the ivories. She unfortunately suffered from epilepsy. It was her opinion that the sheer influx of some colours, flooded her brain with so much electrical activity, that she collapsed into what she termed a mental orgasm of riotous shades. Eventually she ended up in a mental asylum because of it. I used to suffer from migraines for the same reason. Particular sounds would overwhelm me and the inside of my head would reflect Disney's 'Fantasia' - flashes of red, yellow, magenta etc.

Cousin Tom used his synaesthesia in a constructive way. 'The Great Tombola!' as he was known. He could memorise telephone books as a child - not that anybody particularly wanted him to but that was his party trick each Christmas. He was even better at doing mental arithmetic than Uncle Dick but the latter didn't care too much, saying he knew several Mathematicians who couldn't count their blessings, let alone the change in their pocket.

'Ah, here we are! My destination or at least the one before I go shopping, proper - 'The Last Gasp Cafe' as I call it.

'Hello Angela!'
'Hello yourself'
'Good Christmas?'
'Good? Frozen turkey, frozen water pipes and frozen out by his relatives again.'
'New Year?'
'Yes I had one'
'But was it any good?'
'What do you think - the parents were still there'
'Whose parents?'
'Whose do you think? Certainly not mine.
When are you going back to the funny farm?'
'When it becomes funny ha-ha and not funny peculiar'
There was a pause.
'Talking of peculiar, don't you like Theakstone's 'Old Peculiar' or something similar?'

'Marston's Pedigree'

'Begins with a 'P' doesn't it?'

'Yes and talking of pees, I need one before I have a coffee - it's freezing outside'

And with that he made his way to the toilet.

'Angela,' he said to himself, 'has a nice lemony flavour to it, with a tinge of Apricot'

'Hi Mike!' said a figure coming out of the loo.

'Hi Ken!' he replied (That's a nice walnutty name isn't it - perfect for this time of year, just like Chris, with its chestnutty feel).

Five minutes later and he's out enjoying his nice, hot coffee.

'They say we're sensory amalgams but society itself is an amalgam of different inputs and outputs: Different cultures, different languages, different races - forming and reforming into new combinations, ensuring the human race stays stimulated, alive, aware and awake, through this mixing, instead of being bored to death by unchanging sameness; defended by the old minded, to keep things old - that is ensure the established, stays established and inherited control is passed on.

Words evoke physical responses in everybody as with Pavlov's Dogs and conditioning but this is self-taught repulsion and attraction. Other people don't see it as being the same thing as our confused sensory orientation, of subjective-objective unity but it is. It's all association as with intelligence and creativity - only making connections most

24

others wouldn't.

It's funny but this cappuccino reminds me of pink blancmange for some reason. Angela is a nice enough woman but it's Mandy, who I usually come in here to see.

Mandy is sweet.
Mandy is kind.
And with a body like that
She drives me out of my mind!'

He coughed in a nervous way as though embarrassed at what he'd just said - which he was. If only his mother could hear him now.

'Oh well, time to go through my list and actually do what I came here to do.'
And with that, he was gone in a flash of light and maybe just a hint of minty surprise.

ABDUCTION BLUES

They're coming for me - I know it! I can feel it in every fibre of my being. It's those eyes - they're staring at me with intent again, from wherever they really exist.

Nobody believes me of course. This is Britain. This kind of thing doesn't happen here. Even the psychiatrist doesn't believe me (paranoid delusions, in his opinion) but at least he listens - that's what he's paid for. I don't want to go back inside, so I'm careful what I say nowadays and who to. It's alright talking about it on UFO forums (Thank God for the internet!) but in public, no way.

People think it's only in your dreams - if only they knew! You can't get away from them; they are with you when you sleep; they are with you when awake, continually monitoring your reactions, continually feeding your mind with ideas. They feed off your emotions. They measure your reactions. They tell you how 'special' you are. They promise you everything and deliver nothing: You will rule the world one day, they tell you - then at night, oh my God at night...

I'm not sure which is worse - being treated like an animal, a piece of meat at night or being mentally manipulated during the day. Go there. Do this. Meet this person. And you do it, like some living psychic nightmare, in which you are hypnotised to obey, to believe. To start relationships that tear you apart emotionally as at night they tear your body apart, just as callously. Is there no way out of this living hell? Some experiencers say you never escape, even after death. Some contactees say you can't. Like Shamen, they believe death is not the end and that this is some grotesque initiation ceremony - a cosmic joke, if only you could see the funny side of it: Sorry I can't.

And what of the future? Isn't that what hell is? A never ending story of perdition, personified - where you are puppets of a greater force than you are.

Where shall I begin? Where did 'it' begin? I always assumed it was something that happened at puberty, when the sexual humiliation began. It felt as sexy as a cow being milked. It was like being in a hospital - row upon row of shuffling naked humanity, going towards white tables, like moulded plastic, which seemed to grow out of the floor like toadstools but not separate from the floor they were on.

Some struggled against it, being floated to the tables as we had all been floated to wherever it was we were now. Then the four fingered hand would go on the brow and those large, black, gecko eyes would take over, emptying the mind of everything but a sense of peace and blissful emptiness. Slowly the process would begin, of sucking sperm from the paralysed body or harvesting eggs from the female.

'Beats working for a living' one man quipped, mind to mind.
'How can you think like that!' I thought back at him.
'Take it easy. What can you do about it anyway?'
He was right of course. What could we do? Victims all. Could this be where ritual and sexual abuse came from

originally?

The minor operations to put implants in, I could stand, even though it marked us out as 'owned' - traceable cattle, they could monitor and track. The gut wrenching surgical processes were something else: Is this what it is like in an operating theatre; only without the luxury of being unconscious, if they pull out your eyes, cut open your stomach and pull out your internal organs or scoop out your brain, like a soft boiled egg? Now I know how my dog felt, when I callously tossed it into the vets, to be done because that is what I wanted. I can still see the look of terror on her face. Is this what Auschwitz was like?

'What am I doing here? What's going on? Why have you abandoned me, my supposedly loving master?'

Why indeed? I'd never do it again, not now.

According to shamanism, this is symbolic dismemberment. Symbolic!?! They should be here! Still, maybe they were...

Sometimes I'd remember what happened - other times I wouldn't. I'd be asleep in bed and it was like I was a baby again and my mother was picking me up from my cot but I was actually floating out of bed and drifting through the closed window, like it wasn't there. I'd tell myself it was just a dream but I couldn't do that when I found myself outside, after the event was over cold,shivering, naked or partially clothed. My mother told me once that I often sleep walked as a child and she'd find me asleep in the morning, in all kinds of weird places in the house or out buildings. That she could understand but was mystified by where I got the strange clothes, I sometimes was dressed in, when she found me.

'That's not your brother's or your father's - it looks foreign to me, not British.'

At this point the childhood memories started to flood back. The meetings in the woods, after school with 'them.' The other strange kids that appeared out of nowhere, to join me there. The lessons we were taught were about reality and our lives to come, our place in the universe and much more besides...

In adulthood these episodes equated with missing time. I'd find myself in strange places 'after' and have no idea how I got there or look at my watch, thinking two minutes had passed, when in fact it was two hours. Sometimes though it was like being set up to 'be' somewhere specific or meet someone 'special.' I'd just have this irresistible impulse to go somewhere and when I'd arrived it would just disappear. I'd see or find what I was sent there for or meet someone, who'd tell me something 'I needed to know.' All very cloak and dagger. Sometimes it would be a book I needed to read. More maddening were the cruel relationships, set up for you. They'd throw you in at the deep end, with some girl you'd never seen before, then split you up and spit you out as though it a never happened. To me at least, this emotional trauma was worse than anything they did to me physically but both forms of abuse left you scarred. At least with arranged marriages, you get to stay with he other person for life. This was like having a baby, then having it wrenched away from you and yes, they did that to the offspring they created. Too women have told me this same, horrific story, for it not to be true. Sometimes they let them keep the results of these unions but it was always for their reasons, not as an act of mercy. Occasionally these implanted children, didn't come to fruition, thankfully...

I always knew when they were coming for me because the blinding headaches would start again. It was the pressure of their presence. When they left, I'd feel light headed and nauseous or 'travel sick' as I'd like to joke, to those in the know (They didn't always need us there physically, so connected with our minds instead or abducted our astral bodies - hence the joke). We were in a little group of geeky freaks that used to meet in the pub every Friday, to talk about UFOs and all things related.

'Well Ted, what's new?'

'Nothing much - chap in Northwich claims to have seen some kind of strange light in the sky.'

'Here we go again!' said Mark, our official sceptic (Isn't it funny how we moderate our own views, to fit in with our friends or others we feel, we have to kow-tow to because they have something we feel, we need? The party line isn't reality as we experience it but shutting our mouths and lying to please the vanity of our individual pushers and pimps: We feed their egos and they give us everything from sex to drugs, on the obvious side and cars, cash, houses on the more subtle level).

'So Theodore, what was so unusual about this light?' I asked.

'It turned from red to green and passed go?' Enquired Mark.

'No, you idiot. The guy who saw it said it was just slowly drifting across the sky, when it suddenly stopped, so he thought it was probably a helicopter. Then it just went straight down,like someone holding a billiard ball and letting go of it, all of a sudden.'

'Wow!' said Jim.

'That's not the best bit.'

'What was?' Mike asked.

'Well, when it hit the tree line, it started heading straight for him, like it knew he was watching it.'

'Then what?' I asked.

'I turned off the TV and came down here to see you lot.'

'But you said-' reacted naive Jim.

'I lied!' Enthused Ted.

'No man from Northwich?' I piped in.

'But this is where it gets really interesting. When I got outside the back door, there was this bright light in the sky, which suddenly dropped when I spotted it.'

'Core!' said Jim.

'It didn't start coming towards you, did it?' I queried suspiciously.

'You've heard it before?' queried Ted back.

'You bloody liar!' said Mark

'Yeh but it had you going!'

'Whose round is it?' asked Mike.

'It ought to be yours!' I glared at Ted.

'I got the last lot because I knew what was coming.'

'Sod!' said somebody, who should have been Jim but wasn't.'

'I'll get it' I stated and got up.

Just another typical night at the local UFO Spotters club.

My only interest outside this whole paranormal scene was science fiction and even that seemed connected because it was speculative and futuristic. Also every time a Dalek blurted out 'You will obey!' I couldn't help but think of my bald headed friends (or maybe that should be 'fiends'). Funnily enough Mark, our resident sceptic, was also our keeper of factoids. He once told me that there was a correlation between Dr Who series appearances on TV and UFO waves - in Britain at least (I wonder if that is also true of Star Trek in America?).

While I'm willing to talk to the group about some of the things that have happened to me, I won't even mention the stranger aspects to the various internet forums I visit because they are too weird and it simply embarrasses me. I tried it on one site and was immediately ostracised by the members and banned just as mysteriously. It was like my one and only daylight sighting. There it was, this gigantic mothership, floating slowly over the town and only I could see it (nobody around me acknowledged its existence and even I didn't draw attention to it). Years later I would hear it described to a tee as Somebody Else's Problem or SEP. It was like times I went into a trance state and stared blankly into space, when they contacted me telepathically. To those people in the street I suppose, I'd appear to be looking intently at nothing for simply ages. The strange thing was though that apart from me and my mates, the usually busy shopping centre was bizarrely empty when such incidents occurred.

Other times the weirdness got really weird. I'd meet people I knew to be dead and even stranger, someone I didn't know was dead until much later, even though the person told meat the time that they were dead (I of course didn't believe them). Then of course there were the prophetic dreams and surreal episodes - like the time an alien stretched its leg right across to the other side of the road and tripped me up as I was trying to run away (It was night but I was still fully conscious). This was real life screaming at you that existence was really a cartoon.

They told me things about myself that shattered my illusions about who I thought I was. It was the unknown tapping me on the shoulder and telling me that reality was greater than I could imagine. I was shown visions of what was to come and told what had been. They further told me that free will could change all this. I was further made aware of my mission on Earth and that of others like me. We are to wake Man up and drag him screaming and kicking, up to the next level of existence. We're here to rescue him from the self-indulgent and egotistical dead end, that will bring down this society, with its purposeless, pointlessness and general lack of direction.

We are the new shamen, for a new world, with magical powers and high intelligence - armed with a new vision of reality, aimed at inspiring others. We are part of a chain to the stars, consisting of hybrids and genetically altered human beings that will lead humanity off this planet and up to heaven, to join those created us, after we have done the work to save ourselves and indeed this planet.

They told me 'You are sensitive to everything and psychic because you have lived many lives before, refining who you are and what you can do as a being. Those stuck in their own personal hells are by contrast, paradoxically old and new souls - old because they've shut out the light of fresh experience and new because they haven't been born into who they could be and out of who they think they are, now.

Heaven and hell are just directions that lead you further into or out of yourselves. The future leads you up and out, freeing you from the trap of identity that is the past or down and in towards thought. To be reborn is to not know yourself or anything else but be lost in the mystery of existence again. It is the seeking of experience and the maintaining of freedom to be yourself, rather than what others want you to be or expect you to become. You as spirit are potential, just as you as matter are actual. No journey really takes you anywhere - only over the border from fact to fantasy (or fallacy as some would name it). Remember, all you see inside can become real outside by turning dreams into reality, through belief and effort grounding them. All remains pipe dreams until you act, gain co-operation with others and realize that nothing really matters: It's not the success or failure that matters but trying and having fun doing it. Children know life is a game, a fantasy - adults forget because they get buried in the wreckage of their past lives too much.

THAT'S THE STATE WE'RE IN!
A collection of Short Stories

Go forth and live your life. Remember it's what's inside that matters, not outside appearance.'

MESSAGE TO THE YOUNG

If you survive your impetuous youth, to reach my age, you'll have all of the following to look forward to and more (Old age doesn't come empty handed as they say in Scotland): Indigestion, heartburn, piles and a bladder that controls you - not you, it. You'll cut and bruise easily too. Your teeth will be sensitive to both hot and cold things equally. Chewing hard stuff? Just a memory of stamina you've no longer got. Your eyes will run at the least sign of cold weather, until you drown in your tears and your nose may follow suit. Oh yes, you cocky little bastards - no more bare arms and t-shirts in the pouring rain or falling snow flakes. As leonard Cohen sang 'You'll be aching in the places, where you used to play.' One day the wrinkled bag between your legs, will be caught up by the rest of your body and you'll be a dried and crinkly old prune all over.

I hate the lot of you, with your smooth skin and pimply faces. Your ability to run and play in a way now lost to me forever, gets right up my nose but one day I'll have my revenge, if you don't kill yourself first. Old age is the only prejudice that can be guaranteed to catch up with you. The shoe invariably and inevitably fits on the other foot. That cannot be said about race, sex, sexual persuasion or class. We all grow old and one day you'll be laughing on the other side of your face, when the insults you sent out to us, return to haunt you, 'grandpa!'

That poor lad shot in the head at point blank range in Salford recently and that boy stabbed through the heart,

36

crossing a bridge in London a few years back - total strangers, killed for no good reason. They could have grown up to be worth something to society, murdered by jealous scum who were no good to anyone. All you lot talk of is respect, yet you show us none and scant little to anyone else, including yourselves. When you reach my age, who'll look after you because your kids won't. Like you abusing us in care homes now, this is also something for you to look forward to.

I wish you all the corns, callouses and ingrowing toenails I have. All the bad breath and rotten teeth too. The arthritis, the rheumatism, failing eyesight and all the other general aches and pains that come from getting older. The permanently gunged up eyes, ears and nose. The dribbling at both ends (The erectile dysfunction especially, I wish you well with). The veins showing through your thinning and blotched skin - welcome to the real horror of old age! The juddering when drinking, the lack of stamina that leaves you leaning on a wall, on the way home. The deafness, the memory loss, the perpetual dry throat - the old person's smell that hits you, when you reach a certain age (Everything stinks in a different way from that point on). The shakes, the loss of balance as bodily rhythms throw you all over the place, where once you controlled them, right down to the smallest movement.

All your vain wallowing in pride about your appearance - where will that be, when the hair on your head turns grey or falls out and what's up your nose, in your ears and on your eyebrows, turns into a dense thicket? What will you do, when your body turns into a bag of fat or just skin, stretched over bone?

All those smooth skinned, lithe young bodies, having no blemishes except for temporary acne - I envy you! My revenge is knowing that one day you'll end up just like me - old and falling to bits at the seams. All the 'pleasures' you enjoy now, one day will be gone. Sugar will set your teeth on edge, bread will give you heartburn, fat will make you want to throw up. All of life's victories you can say goodbye to because they will simply make you too ill, to enjoy them anymore - the sex, the drink, the cigarettes. All because you've become too sensitive to sensations, to put up with them anymore (Too bright, too noisy, too smelly, too strong tasting). The first half of your life, you're indefatigable - the second, crushed to death by your experiences. The good memories disappear and the bad ones resurface. Do I have a poor memory in reality? No, I just don't want to remember the past anymore or experience the present and as for the future... I've lost all my appetite for life and just want to die. I've watched family and friends go before me and now I'm dead inside too, just waiting for my body to catch up and release me from this living hell of daily life.

I remember how it used to be with my wife and me. We stood together against the gathering storm but now we've fallen apart. We can barely talk to each other. We go through the sham of a happy marriage, like nearly everyone else we know because that is what is expected of you. I remember when we first met and fell in love. We were two strangers, who opened up and trusted each other. The barriers between us dissolved and like blocks of ice, melting into running water, we flowed into one another. Finally finding a mirror of ourselves, we blossomed into one being and abandoned the lies and subterfuge of the world around us. Now the agreed upon barriers have been reforged. We do not let passion push us beyond this. We never meet in no-man's land anymore. Our 'relationship' is constant and consciously maintained because it can never be anything but distant from now on. Love is dead. Mutual exploration is dead. We are dead.

We no longer fight or struggle for understanding but have drifted emotionlessly away from each other. We are polite across the cold spaces between us, going through the senseless formalities but that is all. To outsiders living their own cold, dead lives, we are 'still' in love but robots know the truth, minds know the truth - hearts though won't accept it and die for their treason to hard facts.

Yes, you have all this emotional hell to look forward to and I wouldn't wish it on anybody, not even you - the death of love, the death of friendship, in a crumbling world where we throw ourselves into tasks, to avoid facing this loneliness, this emptiness, we cal our 'lives.' We are battlefield casualties of a war we slid into - one we never consciously chose. Wrecks of yesterday - mind staring at mind, broken heart at broken heart, unwilling to face the pain inside. We pretend that all is 'normal' and act as though it were, in the vain hope it will be again but knowing different in reality.

THAT'S THE STATE WE'RE IN!
A collection of Short Stories

Shut me in my box
Keep me safe and warm.
Shut me in my box
Save me from the storm.

A PAIN IN THE NECK

I've got to get this paperwork finished by tomorrow but all I can hear is this thump, thump, thumping bass above me. I want to get into my head. I 'need' to get into my body, to get this stuff done but no matter where I go in the flat, this incessant noise grabs my attention. I could kill the bloody bastard!

The pills don't work anymore. I've got a splitting headache and I know tomorrow it will have turned into a full blown migraine: Three days in bed, vomiting, sweating like a pig, stuck in delirium...

It starts with tension in the neck. I can sometimes feel a sharp pain where it joins the skull. One day I know this blockage will turn into a stroke that will kill me but there's nothing I can do about it. I tried anger management but left after punching the guy who ran it. He knew how to push my buttons, so I pushed him back. The guy above me is different. I fantasize about taking a sledge hammer to his music centre (Kicking in his door and kicking that thump machine into a million pieces). Any resistance from him and I'd do the same to him. The reality of course is different...We pass in silence or polite greetings. He doesn't care one iota about my feelings, the arrogant sod. It's like being in an abusive relationship. I hate him but my feelings don't touch him. This is a living hell.

I won't let him drive me out of my own flat. Sound smashes your ability to concentrate. It splatters all over the place, so that you become a flat, empty, mindless husk like him. Responsibility is about seeing but he wants to remain blind. Drinking, partying to all hours. Where does he get the money from? How's he even fit for work? Still I never am - mainly because he makes me sick, literally! Why do we have to carry these wasters? Why!? Why!!? Why!!!?

A silent, contemplative society - that would be my dream (A monk to his viking raider). Peace! Peace! Peace! That's all I want! Christ, is that too much to ask1?

I'm a workaholic - I admit it but where would this society be without people like us? They'd just let things fall apart or worse still, rip them actively to shreds. They'd rob people, beat them up, threaten and cheat them, just to get their needs met or expect to be spoon fed like the babies they are. They resent the world and everything in it, including themselves. Work is a four letter word in their vocabulary. Honesty is another swear word to them. They'd rape and kill, to get what they wanted but ask politely? No way! They kill cats, grunt obscenely and drive their cars and lives into the ground because they don't really want to be here. Drugs, alcohol, lack of sleep, reckless abandon of all sorts - anything but settle down and develop inner resources, inner knowledge for dealing with the world they find themselves in. They're damaged goods that they helped damage. Self-sabotage motivates them, not thought. They call people like me boring: Shopkeepers, accountants, academics just as they called the first farmers the same because they were the mighty hunters - posturing and pouting, in their antique roles as time passed them by. They see themselves as predators still and we as their prey. And why not? We've got the real power in society, not them.   They are the victims of this world - dinosaurs left behind when the rest of humanity caught on, to what we were about. We want to see a different world - they are only happy if they can take their old one with them. They want to convert the new world to their old ways, dragging us down into their hell. We don't conquer the outer world and turn it into replicas of our own lives. we conquer ourselves and our primitive fears, to better appreciate anything new and different we encounter. Still enough about them and our efforts to convert them to a better life, raising their consciousness to our level.

I know an attack is coming on because get this vile taste in my mouth that reflects the mental bitterness I feel inside. My skin feels like it's flea infested. I get severe indigestion because I feel life is hard to swallow. Perversely I also get the munchies, where nothing I eat satisfies this craving inside (Stuffing down my feelings because being sick means no longer being able to control them).   Coffee is one of my triggers. I know this from vomiting up a cup I'd drunk half an hour earlier, in a distilled version (no sugar or milk, just the black steaming caffeine).

I reckon the reason migraine is on the increase in this society, is because of the ever increasing pressure on its members.   It's like an orgasm or a massive electrical discharge. Things build up to a climax then explode like a volcano. Epilepsy is that way and I think migraines are no different.

It's like trance dance as seen in voodoo and whirling Dervishes too. Continual motion, leading to inevitable collapse of the organism through adrenal fatigue or society through panic attacks par excellence! I believe were just vacuoles sucking in and blowing out experience or electricity generators, accumulating then discharging energy. I believe too this explains ageing as motion between two points and again dementia as chronic delirium or loss of contact with the world, leading to balance problems, loss of appetite and inability to keep food down (The big trip of unconsciousness as opposed to small deaths on the way). All of this fascinates me and why not? My doctor says I'm talking rubbish in his own particular, polite, professional way. I accept his reaction with a pinch of salt. What does he know? He should be on my side of it.

I awake the following morning. Awake, is that what I really am? I feel like death, staggering about the flat like a zombie. It took ages to drag myself out of bed and look in the mirror. Yes that confirms it - I am dead again. That sallow skin, those lifeless eyes - black around the edges, bloodshot within. Tongue out. Yes it's that white flag of surrender again (another overnight snow storm, covering it). Sometimes it's yellow with vile bile and it tastes bitter and ugly like my mood too.

I just want to curl up and die - oh God, here it comes! The wretch throws himself down before the God of the toilet seat and retches. He prays to the Lord of Vomit. 'Please accept this humble offering - yurp!' Oh God, here we go again. Yurp, yurp, yurp! Nothing there but I don't listen to my stomach. Once more with feeling - yurp!'

I sit there for five minutes, leaning back on my heels. Is it over? Is another eruption on its way? Eventually I stumble back to my feet and walk shakily back to my bed. 'I'm ready for my shot, Mr Romero! No, I don't need anymore make up and I remember my lines perfectly. Groan, grimace, stagger isn't it?' I don't need the thump, thump, thump of his music above me anymore - I've got the recording going on in my head already. Sorry no, make that the sound of blood pounding around somewhere in my crunched cranium.

When I'm like this I hate everyone and everything, including myself. As the old joke goes 'I'm not prejudiced - I hate everyone equally.' I just want to die and want everything else to go as well. Lights are too bright, sounds too loud, smells too strong and touch makes my skin crawl - don't mention taste - even water won't stay down. These dead eyes back away from the world in disgust. They don't want to see, let alone look. The curtains are closed, the blinds are down and the eyes shut against a cold, uncaring world. Every system breaks down under pressure (overload): The final straw on the camel's back or the bucking bronco, kicking back against the heavy load and going into defensive mode (protecting the friend within by closing down and shutting the intrusive enemy out). This is probably why all my orifices dry up or block up, after an attack - gunge in my eyes and nostrils, wax in my ears - skin pores blocked with stale sweat and let's not forget the constipation (if only I could) and the thirty minutes straining on the toilet seat. Everything shutting down and trying to shut out, everything else outside.

I slip between dreams and wakefulness, in a delirium of self-loathing. I spit out the odd syllable oh hatred as I twist and turn, in a sweaty melodrama of contempt. Future? What future? As for the present, I don't want to be here and don't mention the past with all its interwoven failures - the parents I let down, the jobs I dumped for no good reason, the women I discarded like empty cigarette packets, the towns I moved away from. I couldn't face any of it. And now here I am again, wallowing in self-pity and spitting at any help offered me, in a mad rage. I don't want your pity, your helping hand - I just want to be left alone to die! (But I don't die and that is the bitter pill I swallow every time). This hell circles me me like a bunch of vultures , each time this mini-death strikes. I smell and look like a corpse because that is what I am. The world's biggest hangover never goes anywhere - it's always there, waiting in the wings for its next curtain call. Still Scarlett, tomorrow is another day!

Today I feel flushed out like a toilet - drained, empty and my breath stinks like an open sewer. Work tomorrow, if I'm lucky - if not another day in bed at least. The reward of sin is death and I've hit the jackpot again! Another bowl of scorn flakes, to set me up for the day. Hi-ho, hi-ho and it's off to work we go! This anger driven obsession gives way to laughing at its absurdity again and the brave face returns. The sad clown gives way to the mad clown, giggling away at nothing; instead of frowning at something as though this speck of dust, this instant in eternity, was a ten ton rock on his chest (or head more like).

THAT'S THE STATE WE'RE IN!
A collection of Short Stories

Did you know most hospital patients die between three and four in the morning? They call it 'The Death Hour' because of this. I worked out that this is when my migraines start in ernest. Somebody told me that this is because this is the time you run out of energy, nutritionally. I believe this is partially true. It's also when you're most relaxed. When you're awake, your body tenses up naturally, through mental stress and body motion: Move-tense-move (hold in position and let go).

It's like a car going into a garage, to get repaired - you need to decommission the body, to carry out work on it. Take it off the road to deal with wear and tear or the accidents we are all prone to (You can't fix the engine driving along the road, can you?).

One of my biggest insights was when I discovered from a friend that it wasn't the stress that really caused the attack but the relaxation that occurs after, which leads to the body's systems getting flooded, when you let go of the cares of the day. It led to a eureka moment, where stormy weather as a trigger, suddenly made sense. Then I saw a program about mountain sickness and realised it was the same thing too. It told me that mobile decompression chambers were used to treat this condition, in the same way The Bends were fixed in more solid versions of this (pressure problems again). I rushed to my doctor, to ask if the local mountain rescue team had a Gamow Bag I could borrow.

He said 'You're joking of course! I go hillwalking and it's needed for people who can become fatally ill through altitude sickness, not minor headache sufferers.'

47

THAT'S THE STATE WE'RE IN!
A collection of Short Stories

'Could I buy one?'

'Oh yes but they are not commercially available to the general public you know?'

'Why not?'

'Ask the government.'

So that was that.

Looking for alternative treatments, I dipped my toes in a sensory deprivation tank.

In women it can be hormonal they tell me, disappearing in the mid-fifties, when the menopause cuts in. To me it's an emotional brainstorm, which burns out the computer screen, leaving the sufferer unable to function. Lightning strikes can have the same effect - twitching, memory loss, staring blankly into space, like you're in a trance. Everything speeds up - urination, defecation etc. until the body reaches a climax, whereupon it collapses back down again into stillness and the mind into silence. It's almost like manic-depression but physical not mental sensitisation followed by de-sensitisation. Freezing during the day - boiling in bed at night, like an overnight heating storage system. Bitter taste before an attack, sweet one after - like insulin kicking in. In fact I think the whole of life is like one, gigantic migraine attack youthful discharge, followed by collapse into old age's illness and decay: Inability to keep food down, loss of memory and awareness - dementia as the equivalent state of migraine delirium, where you lose contact with the real, solid world and drift away into serial dreaming, contacting dead others, like you.

THAT'S THE STATE WE'RE IN!
A collection of Short Stories

When I first took pills that worked, I could feel my body reacting as the chemical imbalances were corrected. Half an hours sleep and I was right as rain but it was only effective with mild attacks because during severe ones I couldn't keep anything down. Taking showers was something else though. I could feel every drop of water on my skin as though it were hail stones, such was the effect of the mould derivative on my nervous system.

When I went to the dentist, he said I suffered from bruxism. What's that I asked, incredulously? 'Teeth grinding - everybody who has migraines, does it.' he assured me.

'Oh?' I responded.

'Yes, it's the suppressed anger' (as if I didn't know!).

He suggested a gum guard like boxers use, in order to stop it. Worn overnight he said, it would stop the teeth crunching and therefore the pressure in the jaw that led to the headache (That was the theory anyway).I tried it once but it was like trying to go to sleep with an apple in your mouth, so I soon dumped it as impractical and uncomfortable.

Before an attack my skin would crawl as though it was brushing up against cobwebs. My shins got so itchy that I'd scratch them until they bled. I'd also get this funny prickling sensation on my left shoulder and a high pitched whistling in my ears. My doctor said it was probably wheat intolerance. What about all the tension in my back and other joints and the only relief I got was by cracking them? Gas, he said. And the rheumatoid arthritis after an episode? None of this is related to the migraine but is a separate issue, according to him. We'll put you on so-and-so, which should soon sort it out (It didn't, anymore than his answers satisfied my curiosity).

I believe as I said earlier, this proves it's some kind of accumulation/ discharge problem, showing up as the hot/ cold difficulty, body tension and relaxation, plus other polar opposite symptoms: This includes the static I'm prone to, when I touch plastic handrails or pull off my jumper, made of synthetic materials (The spark across my nose is shocking, just shocking!).

I tend to drop things after an attack because I'm not aware I'm holding them. The doctor said its possibly nerve damage or blocked blood vessels as a result of the migraine. This and slurred speech, memory loss etc. indicated signs of minor strokes but not to worry (Apparently all these mini-strokes can build up into a major one in later life, which can kill you and this is the real danger with migraines, even if the attacks disappear in your mid-fifties as with most people: Personally I wish I hadn't found this bit out).

Life is a pain in the neck and then you die, recover, forget all about it, then die again, in an endless cycle of hope and despair. Such is life.

THE SORCERER'S APPRENTICE

Jack had first started noticing the figure several weeks ago. For a start he'd just catch him out of the corner of his eye, then as time went by he'd see him clearer. It was funny but at the beginning he could never look at him directly, like trying to force two magnets together, repulsive pole to repulsive pole. He'd distort and move away but not like an ordinary person would, who was trying to avoid being looked at. It was more like a reflective sheet of silver plastic, being poked in the middle. Even before that Jack would look where he thought he was and see nothing unusual - no dark figure or shape even. He knew though that he was there. He could feel it, sense it but not see it.

Now today though it was different. It was like the whole world had stopped and there was just him and Jack. Their eyes met but this was no romantic meeting across a room. Jack felt probed, prodded by another mind. It was almost palpable. He wanted to take his eyes off this mysterious stranger but knew instinctively not to. It was like he knew that he'd be thrown across the platform and slammed against the far wall, if he broke contact. And then it was over and the being had gone completely. The world unfroze and moved on again, in its daily business as though nothing had happened. Jack got on the train and made his way to work as normal but felt he'd just been stared down by some wild animal.

That night Jack returned home. He had his tea, watched TV and went to bed. Suddenly he awoke in the darkness.

'Who's there?'

He felt a presence - in fact this was what woke him.

'I know you're there - speak!'

Only silence filled the room.

There had been a spate of burglaries in the district, over the last few months, so he feared the worst. He wanted to turn on the bedside lamp but feared any sudden movement might bring down a blunt object on his head, by someone whose eyes had adjusted to the lack of light and who was fully conscious, unlike him.

Courage overcame dread. He made a sudden grab for the switch. The room was empty.

He jumped out of bed, looked under it, behind the curtains, beside the wardrobe. Nothing.

He went into the kitchenette. Nothing. Tried all the doors and windows. Locked. Shut.

He made himself a cup of tea and went back to bed, for an unsettled sleep.

The next day he was at the station again as usual. No sign of the strange being this time.

Suddenly a voice popped into his head.

'Yes it was me last night.'

He looked around, agitated but no-one was close enough to have said anything to him that loudly, without others hearing.

'No, you're not imagining things. I'm really here and speaking to you, mind to mind.'

'Who are you? How do I know this isn't all happening in my head?'

'Look at the man to your immediate right - the one talking to the pretty girl on his left.'

With that Jack turned his head slightly, to get a better view but without being obviously intrusive. The couple were laughing at the man's animated conversation, when suddenly stopped in mid sentence.

'What's wrong Charles?'

He didn't answer but looked furtively about the platform, his body still as ice. Like a rabbit eyed by a fox, he froze on the spot.

'I can do that because he doesn't know me. I've played that trick on other regulars but couldn't do it with them again for this reason. They've become acclimatised should we say.

About my visit last night. When we do this, that is my kind, ordinary people think it is The Devil or demons - incubus, succubus, vampires, ghosts or what have you. They do not understand, so they invent names and roles for us that really have little relevance to what we are. Their fevered imaginations must supply an answer, so that they feel in control and not victims of a power greater than their own. Tragic. Nowadays we even get mistaken for aliens - they exist but again the feeble urge to create an answer comes into play - they don't know, so make up explanations. We are sorcerers, magicians, shamen. We are immortal outsiders, looking in on the boring world of the mundane. The wallowing in materialism and petty, selfish means to ends, disgusts us. Don't get me wrong - once we were like them but have moved on, grown up, abandoned childhood and we would like you to join us and do the same.

You are already hypersensitive and aware. This is the first step. Like waking out of a dream, you will discover who you truly are. Your allergies, migraines and seeing things your peers cannot, should have told you that you were different and didn't really fit in here. The flying saucer you saw in your youth that your friends couldn't or else interpreted as a star, plane or something else mundane, should have told you that. They didn't want to know something greater than themselves existed and defended themselves against it, to protect their tiny egos from being shattered - you didn't. You embraced it whole hog. The incident your father never told you about, where he too saw a UFO by the quarry edge, disregarding the voice in his head trying to tell him it was shed. Like you he could see and now it's time for you to join us.'

'What if I should say no?'

'That is your choice. The addictive lifestyle these mere mortals have, will stay yours too as will all the niggling intolerances and prejudices that drive this world. These are sheep, hooked on effects - is that what you wish for yourself?

I, we, offer you space and time, not herded cattle flocking together in fear of the night. You will be alone - even I will only be with you occasionally. You will no longer feel lonely in a crowd but powerful, strong, yourself - no longer panicked into reacting to an imagined threat. My will could make one of these ants walk off the platform, into the face of an incoming train. They can make us do nothing because they refuse to even acknowledge our existence. I'm invisible - I don't exist to them, except in nightmares and as the voice of conscience in their heads which they ignore or struggle against, driving themselves insane. We are aligned with the will of the universe - they are not.'

The voice stopped. Silence fell into the mind of Jack. He looked around and saw the dark figure again - this time not on the platform opposite but behind him on the stairs. The shape beckoned and Jack followed him into another world, another dimension outside normal time and space.

His last thought was, would anybody notice the difference? (They didn't).

OCD - THE VISITOR

Just when he thought it was safe, Jack had a visitor...

'Don't sit there, that's my seat!'
His friend jumped back up automatically.
'Keep your hair on old son,' said Rich, alarmed and amused almost simultaneously.
'Sorry, sorry Richard. I don't get many visitors nowadays. Sit over there,' he said, motioning to the 'guest' couch.

'How long has it been?' Rich asked, remembering the time they'd been together.
'Two years at least.'
'What have you been doing with yourself, all this time?'
'Oh nothing much. You, yourself?'
'America mostly. Company sent me abroad. Me and Sally eventually tied the knot...two kids...blah, blah.'

Jack's mind remembered the initial ring of the door. The question that shot through his head was who could it be? Not the postman - too late in the day. Hopefully not that old busy body from next door, calling to see if he was alright.
'Now your parents are dead, it must be lonely all on your own.' (No it was wonderful to be alone and not bugged by other making demands on your time, he thought to himself but didn't say).

Thankfully it was Richard. Old school friend and drinking buddy, from the time after that.

'Jack, after all these years!'

A hand reached out to shake his. He didn't grab it, so Richard lunged forward to hug him.

'How are you mate?'

Jack had refused to grasp the hand - it might have been sweaty, dirty or germ infested. The hug took him completely by surprise too. The smell of body odour, cheap deodorant and women's perfume, assailed his sensitive nostrils. Revolting! No wonder he'd shut himself off from human contact as much as possible, over the years. As Richard let go of his hold, the smell of bad breath and stale alcohol, wafted into those sensitive nostrils again. Instinctively Jack held his breath, to keep the worst excesses out.

'I'm not too bad. You heard about the accident and my parents being killed?' (He tried to keep his composure, given the situation but it was hard to put up with this attack upon his senses: Now he knew how Gulliver felt in the land of the giants. Disgusting!).

'Yes, Lolly told me about it or was it Ian?'

So that was how the exchange had gone at the door and now they were in the living room.

'Do you see much of the old crowd?' his friend asked.

'-Excuse me I'm forgetting my manners. Tea? Coffee?'

'Tea please.'

As he sauntered out of the room, he answered Richard's question in retrospect.

'No, not any of them - at least not recently.'

Within a few minutes he was back with a tray, on which resided two cups and a plate of biscuits.

'Don't touch that! That's my cup!' Jack blurted out, almost irately.

'Sorry. I'm just so used to my own company nowadays and things being done in a certain way.' (Like Frank Sinatra, it had to be done his way or no way, down each and every highway).

They talked for ages and then Jack got up, took the tray back into the kitchen and ritually cleaned every item. He rinsed everything first. Then he washed them in boiling hot, soapy water. Finally he rinsed them again - only this time in cold water, so he wouldn't taste the dish washing liquid, next time he ate or drunk from the vessels.

As soon as his friend was gone, he knew he'd have to clean all the door handles and wash his own hands too, to remove the greasy fingermarks and kill the germs that human skin carried.

'Well, I'll be off' And before Jack could move his hand, it had been grasped and shaken, though still wet, in the dangerous manner that all bug infested skin, passed on its disease laden layer of death! (Bath time tonight and serious scrubbing of his palm, once his guest was gone. This was the definite order of the day, his panic filled mind shouted).

'Why can't people leave you alone, to live in peace,' he thought.

'Thank God he hadn't asked-'

'-Oh, can I use your loo before I go?'

'Is the man telepathic?' he thought again.

Jack motioned to the door in the entrance hall.

Somehow he controlled himself, though his natural instinct was to shout 'No!' in a loud and annoyed (frightened) voice.

He listened to the sounds coming from the toilet, tortured by the fear it wasn't just a wee he wanted. The tinkling of liquid hitting the porcelain, greeted his ears. Thank God for that but there was still the chance he'd accidentally peed on the floor. This was never a worry with female guests, should he ever have any but that was that and Richard had gone. Blissfully alone again. Totally king of his own beautifully scrubbed castle, once more - germs and dirt outside, him inside!

ASHBURGER'S SYNDROME

I'm clumsy. I've always been clumsy. People are terrified of telling me to take a break because I do. I break cups, glasses, plates - you name it. They nicknamed me Zorba the Greek, one place I worked because of this. Talking of jobs - with me they've always been few and far between. nothing lasts long. I either get bored and leave or get sacked.

"Now look what you've done!"

or "What happened to that order I gave you, to send out to Mr Harvey on the eighth?" (Well he didn't specify which month, did he?).

To say I was socially inept, is mildly true too. If I had a drink in my hand, I'd either drop it or spill it on someone.

"You clumsy idiot!" (Well yes, I know that - can you be more specific or add something else of interest to that point?).

I was never a great talker and got on better with kids and animals, than I ever did with adults or the human race altogether.

"Stop grimacing at me you nutter!" And other plaudits like this, would come my way. Talk? How could I? I could barely get my body to work, let alone my brain. Occasionally I'd let slip a terrible pun, to break the ice, in social situations. Every time I tried to be clever, an uncoordinated load of stumbling rubbish would come out.

"What do you mean, I-I-I, ig ag ooh?"

Ruthless mickey takers at work or down the pub, would plough right into me as soon as I opened my mouth, so I shut up again or I'd burst into hysterical laughter as I found the joke funnier than anyone else.

"For Christ's sake shut up! The joke's over!"

Then there were the times I couldn't understand what anybody else said. It was like that Far Side cartoon - What you say and what a dog hears:-

"Blah, blah Rover. Rover blah, blah."

It was like I was hearing a foreign language or none at all.

"Cat got your tongue? Well it bloody should have - you don't use it enough, to need it!" (Ha-ha - very funny I thought but couldn't stand the humiliation of trying to actually say it).

Hugh and Milly Asian? Now that's a couple I know well! Yes, my literal sense of humour categorized me as autistic, even if nothing else did. Then there was phonetic spelling.

"That's not how you spell it Wright - get a dictionary!" (Wright, wrong again! School, who needs it? If they want to spell it that way, why can't they say it that way too? It's all so jumbled up and illogical!).

There's some legend that says having Asperger's makes you a mathematical genius - not me. On the way to school I obsessively count the telephone poles, yes but I couldn't add up to save my life or yours, when in the classroom. Oh yes, the stories of us being selfish and self centred are true. We live in our own little world and you can't enter, even with a valid passport. Our borders are closed Mr Schickelgruber and nobody can come through without our express permission, so turn your tanks round and go home.

THAT'S THE STATE WE'RE IN!
A collection of Short Stories

We are a strange mix of contradictions - egotistical, blunt in our speech, when we do open our mouths. Bloody minded and stubborn, yet fearing confrontation because in a fight, we wouldn't know when to stop - at least that is what we believe. It takes a hell of a lot to get us going and just as much to put the brakes on: Quick to anger, slow to cool down and come back into Earth orbit, if we don't miss it altogether. Innocent, vulnerable, trusting and blundering. It is this openness and honesty that turns us into the brainy creature we so often are. While others play about in the classroom or outside it in the playground, developing their social skills through interaction, we shut up, sit still and, look and learn. We shut them out and let the light of understanding in. Ordinary people connect with the outside world, through talk and physical contact - not us. We are geeky, clumsy and inappropriate in our comments and movements but we connect internally with ideas. They can dance, play football, cake on make up or make cup cakes but not us. Books are our only friends - failing that our computer screens are. We'd rather text than talk, write and read rather than speak - even to each other. We want to know how the universe works and maybe even one day, we'll find out how we work but not today, oh no, not today...

We know we are not liked - even feared and despised by some people or why attack us? (You only bully what you're afraid of - what challenges you to be what you are not or at least makes you think about it as a subject). Limited intelligence, criminality and defensiveness go together - leading to ignorance and suppression, by those wanting to shut out the light. Perpetual motion and emotion, keeps them on the move but not us. We don't want to leave home or even go out. We just want to collect our train numbers or plonk about on our computers in peace. Failing that we want to vegetate in front of the goggle box. We are not active participants in life. We are just passive viewers, along for the ride (Don't ask us to drive -we're not up to it or up for it either). We understand sound and motion go together (as with music and dance) but we are detached because we are observers of life, not activists (We don't move with the times because we are lumps of rock - orderly and controlled, not relaxed). We see only chaos and confusion in the world - danger we are not ready to face. Go for a swim? No thanks! You could drown and then there's all the pollution in the sea and God knows what in the rivers and swimming pools! We don't enjoy our lives, we study them for that great examination in the sky, when we all kick the bucket (Did we do well?). Live our lives? Maybe next time. Spontaneity is for wimps - we love routine. Order and discipline, that's us.

We're not in our bodies but always outside, looking in. This explains our odd gait as we're not in contact with life or society's natural rhythms.

We feel continually under stress because we are. Our twitching, tics and odd mannerisms show this. I need to crack my joints continually because of this (neck, between the shoulders, lower back (especially this point), ankles, knee caps, wrists, fingers and toes - by the way did I mention we're obsessive list makers?). This is why you'll see me and others like me, suddenly tilt their heads to one side or move our hands and feet into strange positions, for no apparent reason -

we need to relieve our spasticity (Perhaps this is where 'Jerk' comes from as an insult?). It could explain the difficulty swallowing, indigestion, sensory sensitivity and allergies as well. Maybe too, it explains the dietary fads of eating nothing but a particular food - like crisps, beans, bread or biscuits, for months, even years on end (I've heard that we're carb eaters, avoiding protein and choking on fats).

Is it any wonder that we're stressed? Our attention to detail driven characters, fear of making mistakes, rigid personalities (love of tight clothing), passion for order, discipline and routine - all contribute to the pressure we feel under and put ourselves under. If we weren't so visually orientated, we probably wouldn't be so language impaired, continually swallowing nervously in social situations. This passivity and receptivity is probably what allows us to be so logical but it also leads to the need for space and the temper tantrums that follow, should we not get it and find we cannot cope: The sensory influx that drives us insane - the obsessive compulsion to wash our hands and protect ourselves from every other potential danger , turns us into an explosive powder keg of emotions, which blows up like a volcano every so often.

They say it is a male thing - this turning down and in, in curiosity, then up and out with answers and insights. This mental pressure is the same as physical pressure as in sex and other expressions, I believe. The physics of it is male concentration versus female dispersal of attention and energy. This is why males are more volatile and suspicious because of it (wound up and easily triggered into action, rather than relaxed and patient). Personally I think that giving birth is the true female orgasm and I read recently of a doctor who had the same idea, so I'm in good company. In both cases, sex and birth, lead to release and relief, with post natal depression being the equivalent of a man dropping off to sleep after the event but on a bigger scale - physically and emotionally drained.

I often wondered if my lower back flexing, was down to some kind of static build up, needing to be discharged through sex or movement of some kind. Maybe it builds up through hip and spine motion? Kundalini eat your heart out! Maybe death is the biggest orgasm there is and that is what we're dying to find out?

Am I obsessed with me? Yes but I am you. This is why I have such an identity crisis too. I know the world in general but not me in particular. I'm a chameleon that blends into the background - a Zelig like character that is invisible to all, disappearing before anyone noticed he was there: Death, where is thy sting? Toaster, where is thy ping? Tucking into my three slices of burnt bread (never four, never two), I slide into my life of dull obscurity and wish you goodbye as you drift on down the river of life and I stay stuck on my island of sanctity, worshipping existence in my own monkish way, trying not to make a habit of it and failing...

URBAN SHAMAN

I am an urban shaman. You would not know me, even if you saw me. We slide between the two worlds of sleep and wakefulness. We are inconspicuous and need to be, to go about our work. We are the existential back room boys (and girls). You would pass us in the street without a second glance. You might throw us money, thinking we were tramps or shun us for the same reason. We would accept the cash - not because we want or need it but simply to avoid the attention that any kind of conflict brings. We might provide impromptu magic shows for the odd change or even for the fun of it but most of the time we do not want to be noticed as only the free can travel and study the great mystery of life.

You, weighed down by ordinary life, are trapped in the here and now. We slide between heaven and Earth, and oh yes, hell when we need to. No fast cars, fast women, fast life styles for us. No families, no friends - just colleagues. No home, no possessions. No past, no future. No recreational drugs, no alcohol - what we do, we do for a purpose. No careers, no jobs except for passing needs.

We shun your queries, your stares. We ignore your comments, your attempts to distract us and show off. We are passing clouds, the wind that disturbs you as it passes by on its way somewhere else. We are the strange figures in your dream landscapes - that which you see out of the corner of your eye and question 'Was that a person?' disappearing when you look at us directly.

THAT'S THE STATE WE'RE IN!
A collection of Short Stories

We have the honed willpower you call hypnosis, which your scientists ignore, thinking the world works totally by accident! Pah! What strange beliefs! The only accidents I know of are where people abandon control of their lives end up in a mess because of it. We seize our lives from the instant we enter this world as conscious beings. We don't claim birth as an accident but as an intention to get born here, now. The only reason we fail in any task is that we allow ourselves to be distracted by outside forces - not the shaman though. Of course, if you don't want to take responsibility for the world, it's easy to claim you can't do anything about anything else. Poor victims, needing somebody to rescue them and wipe their childish behinds! (We do occasionally save people but it is more a reminding them of who they are and then letting them grow again, by letting them go). We control our world - it doesn't control us. This doesn't mean we force it to do something, just move with it in harmony, on the great journey that goes nowhere.

When others come to us for help, we usually treat them with disdain. We ignore them, telling to be on their way. The weaker ones go because they cannot resist our will - the stronger ones push back against our hypnotic stare and stay. Then of course we can make ourselves invisible or vanish into the crowd - shape-shift or blend into the background, if people become really tiresome. The good ones see through our tricks though, chasing our energy or seeing it still in front of them, even if in a different form or none at all.

"Go passed me!"

"Ignore me - I'm not really here."

But they don't. We read their minds, to find weaknesses

- then throw these in their faces, like Moses changing his staff into a snake. We watch the weaker ones run, with amused smiles to ourselves, while bystanders act bemused as they 'usually' see nothing (telepathy is a wonderful thing at times). The really stubborn ones follow us, when we slide between realities and we know we have them, like a fisherman playing a fish. Those ones become our chelas, our pupils, our followers - until they learn to follow themselves (stand on their own two feet).

When they want to get away but can't because they are stuck to us like tar babies, they resent us. Do they resent food though, when they are hungry? No! So why this act? It is just the childish attitude of those in this civilisation, who don't want to owe anybody else anything and who don't see that at this stage because they can't. When they do, they show gratitude for life and move on, realising that this attitude alone traps them or 'appears to: We have opponents in our world but not real enemies. Death (time) is that which we treat with respect, knowing it chases us all and changes us all.

We look with humility upon the lost - those caught in the illusions of their own being. These you would call mad in The West but are considered blessed in The East. They of course consider themselves to be God or chosen by God. We know this to be true by the glow of their auras. It is just unfortunate that they are not anchored in the mundane world. They are like flies to its spider's web. Death tries to hold them in this world as sacrificial lambs. We try our best to rescue them, before the authorities try to convert them or drag them off the streets as they do all non-conformists, who might disrupt the system. The comprehender's try to make this conversion as painless as possible, to the degree the system allows. We drag them out and do the same, until the fever passes. We nurse them back to normality, change their nappies and let them wake up to where they are physically.

We see them revelling in the significance of all they see, which they take as God speaking to them, personally (It is not the message that is wrong but the interpretation). We are not angels or your own personal Jesus: Nobody comes to God through you alone. This power is for us all, individually and collectively. Some reach it through drugs - others through meditation, natural talent etc. Some, like us, do it through hard work and discipline. The multiverse is there for us all. Walk wisely wherever you go - you're walking on yourself, remember.

For us relationships, when they occur, can be physically powerful and if they fail, emotionally devastating to a degree that no ordinary person could stand: When we go insane with grief, the mountains shudder, the heavens open up and the sky fills with our rage and pain. Our commitment to the world, the future, makes us feel things differently and in such depth because this power leaves us open and hypersensitive. Ordinary, normal people carry such character armour that a piano could fall on their head without leaving a dent. We don't have that luxury. A grain of sand hitting us in the face, makes us jump as though it were a brick. We know where and when things will happen, so also know to be there or not, depending upon what is coming. Do we warn people when it's bad news? When we know they will listen and who exactly it is will listen. Mainly we say nothing and drift away, like animals before an earthquake - demonstrating by example. Most people are too reasonable, to pay attention or quote facts us at us as if either will change what is happening. Others want to burn us as witches for trying to save them. The sensible ones pack a few essentials and depart as silently as we do (the scared don't even do that). They see the truth in our eyes and run, not even looking back. We see the world today as it is and warn who we can but the shell on most people's egg/ ego is simply to thick to crack open and let the light of truth in. Their lives are just too guarded against change.

We are the world - the world is us but we must open up like seeds, mature and grow, to become part of it. All things must come to an end. this includes us and the civilisation we're part of. The signs are there for all those with eyes to see. The cohesion is breaking down, the seems are splitting and we are all going to be spilled into the void.

We cannot rescue you anymore. We cannot contain you, my children. Like a dam, this society is going to burst its banks and swamp the valley below. No time for flood warnings - The Titanic is sinking - launch the lifeboats and pray to God we all come through this latest threat to humanity, okay.

Is this truly the end? No, only a 'see-change' but not all are aware of it.

'Roll up, roll up - the next show is about to start! Take your seats for the future and kiss your current identity goodbye!'

Some people are intimidated by us - others, spooked by our intense stare.

'We look right through the acts of men' as Shakespeare so eloquently put it. Human beings are open books to us - children whose games amuse us but don't fool us. We wish they'd grow up and when they don't, like all good parents we have to remind them who's the boss and what the rules are (unpleasant for all of us as we don't like dishing out punishment, even if it's merited as it interferes with the prime purpose of life, which is growth through the mechanism of being free to chose your own route; necessity though means doing what ensures survival, even if unpopular). They hate us stepping in and call us evil because of this curtailment of their suicidal stupidity.

They call us effeminate or mistake us for homosexuals because we stand between the two worlds or the male and female paths. Sitting on the fence, we see both sides of the argument and rule with this in mind as olden day kings did once (Think, Lord of the Rings and King Arthur - we are society's Merlin or Gandalf, leading and advising where we can, in the hope we can help you avoid too many mistakes or heartache). We cannot live your life for you but we can caution you about the possible consequences of your actions.

We find you childish and pretentious in the extreme but then maybe God thinks of us as being young upstarts too? You try to teach us to suck eggs and act like we don't understand your silly symbolic language or in jokes. Grow up! We have seen more mature and ethical monkeys than you. You sabotage your own lives and think that is being clever. If you don't want to be here, that's fine but stop lying to yourself and blaming others for your unwillingness to put effort into your own lives. Go away or stay, if that's what you really you want. All the Great Spirit asks of you is that you be honest and if you cannot give that, then your entry ticket is null and void, and you will die and leave anyway. We created a world for you and how do you thank us? By treating us with contempt and the planet with disrespect. What future is there in such an attitude, for yourself, if no-one else?

We are like Captain Kirk in Star Trek 3, The Search for Spock. Our patience is nearly ended but we won't push you into the chasm after offering you the hand of friendship. What we will do though is turn our back on you and do nothing, while you flounder about, trying to save yourselves.

We were here, long before your race evolved. We have had many forms and lived many lives. We are born wise babies and die even wiser old men. We are the future and the dim and distant past. When we awoke, we stayed awake forever - witness to all of God's creation. We are The Watchers - lurkers on the threshold of this world and that one which is to come. We desperately want you to awake too and join us in the lonely night, so that we can all go forward together, for all our sakes. The world is a wonderful place - add to it with your presence. We cannot force you to stay, only encourage you to voluntarily remain. Will you join us in this joyful pursuit that keeps us forever young at heart or fester in a mind that hides from the truth and justifies its failures to exist, through logic and reasoning powers?

The dawn is approaching. We must disappear out of your dreams and blend into the shadows again. You will not recognise us, unless you come consciously looking for us during the day, for like vampires, we are creatures of the night mostly as all old things are. The power of the dark beckons you to sleep and join us as light calls out to the young. Death and rebirth - neither can be avoided. The endless repetition wearies us. This is the price of immortality - knowing that the merry-go-round will last forever, dropping us off at the same point, eternally.

We yearn for death in a way you never do. We know every blade of grass, every stone, every speck of dust here because every return buries us deeper and deeper in awareness of the planet we inhabit. We are your rocks, your certainties and you for your part, bore us to death with your predictability. This is not your fault. We are your bridgehead to a new world but we'd prefer to be in that position ourselves, wallowing in oblivion, full of joyous discovery, uncertain at every step. The past is ours, the future yours - enjoy your childhood.

JUST ANOTHER GROUNDHOG DAY

Well, here I am - looking out the same window, on the same world as I do every day of my life and surprise, surprise it's still the same old crap I see. The same old people going off to do the same dreary old things as they do every day of their lives.

They say that I'm an unkempt old sod, who doesn't give a damn about his appearance. They're right - why should I? In fact why should I care about anything, when nothing I do matters anyway and nothing I do changes anything.

I used to be married but 'she' walked out on me, several years ago. I found out that every argument, every fight we had was just another nail in the coffin of our relationship. I watched the light go out of her eyes and felt it go out of mine. We talked at each other in the end, instead of to each other. We no longer fought tooth and nail. We didn't talk - full stop. In the end it was better she left as staying was killing us both - now, I'm just dying on my own, buried in my room. Every day has just become an endless repetition of every other day of my life, all blurring into one. Since losing Susan and the job, I've had nothing to either inspire or rile me. Before it all went wrong, every day was an adventure to look forward to - a series of changes, keeping you awake and alive, ready for the next one.

My home has now become a prison and life a dead end, in which I'm gently fading into the background of my own life. The door shuts and we die inside yet again but this isn't fiction, this is real life - our ugly own, not some distant possibility but present truth.

Why kill yourself, when you're already dead? (Oh yes I thought about it). Those without life cannot be bothered to even creep out of their graves. They rot and moulder in their own filth and filthy, corrupt minds as I do (Even a zombie has some spirit, driving it to escape it's rotting shell but not me). Those with even a half decent spark, run from death's cloying helplessness but not the apathetic, like me. We never run wild, never imagine, never escape into anything new. We wallow in the past because we cannot be bothered to climb out of the hole we've dug for ourselves.

I think about how things could have been occasionally but know I couldn't face anymore pain, to get it or regain my toe hold on life. This I believe is the fate of the old - to have your dreams shattered on the reefs of despair - to give up all hope and all forward progress, sliding into despondency in its place. Lewis Carroll said that you needed to keep running to stay in the same place. Well I've stopped moving. The world continues to spin, taking everybody else away from me. I stand and watch as the noisy hullabaloo disappears into the distance, without me.

I used to have a wonderful memory and a big vocabulary. I was captain of the local pub quiz team. It takes effort and concentration, to collect your thoughts and project yourself out into the world - to remember all the junk you've picked up and filled your head with over the years, to take care that you're not repeating yourself, so that you don't. I mostly use single syllable words nowadays for this reason - it's easier to remember and who cares if you use the same words twice in a row? I'm like a footballer, who could dribble as imaginatively as Jimmy Hendrix could play the guitar but not now, no not now...

Alzheimer's? Dementia? No, just depression. The older we get, the more bad memories we accumulate - ones we'd rather forget. All the good ones exist in the past - hence we drown in nostalgia. Do they see the world as it is - the happy-clappy brigade? No, they see happier times and run from the present, projecting over this world a film that that is more pleasant than the one currently playing, in this packed theatre of realism.

My mate Duggie had it. When I first visited him in the care home, he was a little doddery on his feet but fully compos mentis. Then the doubt started to creep in as we recounted happier times and happier crimes. Finally, I went in one day and his body was still there but his mind had gone.

'Hello Duggie!' I said as cheerfully as I could but it rang as hollow as the figure in front of me. Within a year, even that had gone. The funeral was the last time I saw his wife, Ethel. I wanted to speak but the gloom of the occasion and the gloom of the tear filled sky, stayed my hand and I said nothing, did nothing. Cold coffee, a piece of cake, a few words with old friends and relatives of his, and it was over. He fell silent into his grave and I returned to mine.

The dust lays as thick as The Sahara here, only disturbed by the odd fly, seeking its next meal. The curtains in an earlier time would be rotten but these man-made fibres mean they last forever but fade with time.

The clues that helped me distinguish one day from another, one action from another, have gone. I can no longer tell if what I'm doing is for the first time today or the second. All time has become an undifferentiated mess of sameness. Words I would have carefully crafted and slotted into a sentence, come out in no particular order and make no particular sense. Oh Duggie, am I heading for a stay in a 'couldn't-care-less home,' like you? I hope not but then maybe,when you get to that stage, you cared less, even than I do now?

Am I angry at losing my job? Am I happy to be stabbed in the back after thirty years of loyal service? I'd have to have been, even crazier than I am now, not to give a damn, wouldn't I? Do I hate the bastard who sacked me? No. It took me years to get over the pain and humiliation but now I feel nothing and haven't for some time. To feel pain and anger is to hold out some vestige of hope for redemption or revenge but neither sting me to life anymore. At first I wanted revenge, oh yes but I picked on the wrong person - me. I beat myself up and looked for excuses to beat other people instead. I glared and snarled, and they avoided my gaze. That look of sheer hatred, poured out of my eyes, like blood being squeezed out of a stone. Nobody would talk to me and who could blame them? People would cross the road or look down as they passed me, to avoid looking at my face and risk confronting the madman. I was just itching for a fight and a chance to let all that hurt pour out onto others.

THAT'S THE STATE WE'RE IN!
A collection of Short Stories

When I looked at them, all I saw was happy smiling faces that I interpreted as grimaces and sneering. Every overheard conversation became full of snide comments about me, even if it wasn't in reality. Even Susan wouldn't look me in the face and that was the start of the bust up of our marriage. Like the prisoners in The Maze Prison, I went on a dirty strike, in protest against an unfair life as I saw it. I didn't shave. I didn't wash. I refused to join the human race that had betrayed me. It could get lost for all I care! (Except it was me that was lost, going nowhere.

What motive do I have for doing anything? I'm no longer preparing anything anymore, for anyone. My job used to keep me on a permanent high as every new day bought a new challenge, every minute a new difficulty to be resolved. When I used to care, I used to feel - now nothing and no-one touches me. I make mistakes and it means nothing to me...but then, I'm nobody to anybody nowadays, anyway. I have no power,no status. When you want to live, you want others to live; when you want to die, you want others to die too - this is the real dark side of the force. When I was younger I wrote a poem that went

Hold on world,
hold on please
- I'm begging you now
on bended knees

but that doesn't apply anymore. I'd like to be polite and tell you what I really think but I don't do polite anymore, so won't say anything at all. Sartre said hell is other people and he was right, which is why I'm on my own nowadays. Why do we want to be heroes? Because we know we're villains (losers - frustrated by what we don't have, know or can do). Enlightenment is simply that step, where we go from helpless failure to knowing winners: Understanding dawns and successful action follows - knowing what the answer to our problems is (The only thing that we ever truly fight is our own ignorance and it's this that drives us mad with despair at our own helplessness). We hate the isolation and loneliness of failure but love the adulation and recognition that success brings. I of course have none of this in my life, so I'm a wise failure.

Funnily enough I was told that time drags you're depressed but I've found it's the opposite. When the gap between waking and sleeping is filled with nothing, there is nothing to distinguish any part of the day from any other part of it. I have no details to latch onto as they don't exist in my world./ It's simply the switching on and off of a light (or maybe that should be 'life'). Existence is like sand slipping through your fingers - nothing solid to grasp. It's like being drunk but not having a hangover. The same confusion, the same clumsy disregard for everything and everyone else.

My doctor sent me to a psychologist, when I was first diagnosed. The hospital had this guy in it, who just sat in this wheelchair, staring into space. The doctor told me he was suffering from clinical depression and I was lucky not to be that bad. I told him about a friend, who'd been in The First Gulf War and was just like him but the medics called it post traumatic stress disorder. Same thing in some ways, he said -

takes you from the world and leaves you shut off, immobilised - similar to catatonia but not caused by a chemical imbalance. Nice man but he could do nothing for me and knew it. It did split up the week though and gave me something to look forward to.

> I look at my wife
> and think about the time..
> Tick-tock, tick-tock...
> When we first met
> and waltzed hand-in-hand,
> through life's doorways...
> Tick-tock, tick-tock...
> Now I stare at the clock
> on the mantlepiece,
> watching my life tick away
> and think nothing at all,
> about nothing at all...
> Tick-tock, tick-tock...

LIFE - A USER'S MANUAL

I walk through life, like it wasn't there (and for me it isn't). I see through the acts of men - no motive confuses me or passes me by. I don't allow myself to get caught up in the games people play - childish and transparent, like all the actions of children. To me they are laughable, so I don't get drawn into them and get hurt as I've seen others do. Innocent? Gullible? Mot me any longer. Life is a book and I've read it from cover to cover. It bores me now where once it excited my curiosity. For me, it is time to move on, get reborn elsewhere, so I want to share what I've learnt with others and with you in particular as my followers - you who thirst for knowledge much as I did, when I was your age. Your age? Maybe it would be truer to say, your level of maturity. They fight like Jason's Bean Men - you have moved beyond that blame, shame, guilt playground, to higher terrain, where you won't drown in such a negative sea of emotions. They are trapped in a way that you have freed yourself from. I offer you a new dream, a new freedom to be...

Life, for your purposes, must be understood as two streams. One takes you down and in, into this world of thought. It isa dark realm of half-glimpsed, half-truths. You however see the world as it is - a whole unit of being. They take the world apart and bury these sections of reality, playing hide and seek with existence. You put the universe back together, to make it work (Heaven leads us up and out, into daylight as hell leads us down and in, into everlasting darkness). The future has no identity - the past is nothing but. They laugh when we say we've found ourselves by abandoning all they consider real and valuable. Materialism and matter is a grave they lose themselves in. Spirituality and abandonment frees us to fly above the world, soaring like eagles, instead of grubbing in the earth, like tomb robbing worms; hoping to gain sustenance from the excrement of all who have gone before and calling it 'gold' (More like fool's gold). Victims are lost in despair because all they can see is their limited world - victors try to rescue them because they can see beyond their limits (The bigger picture, rather than the dot that is 'I').

We are like Gods, who have stepped back from the illusion, staring at the puny little world our compatriots are lost in and we were too, once. They are immersed in the dream, so that cannot see how small the prison is that they have trapped their consciousness in. We are the unknown that taps them on the shoulder, reminding them that there is much more to life, if only they'd turn round and look. We are the conscience of the planet as they have none. They hide from the light as we glory in it. We want to see, know, do more as they want to hide their light under a bushel. They want to hold onto the old, the familiar, the certain as we seek adventure in the new, the different.

The two streams I mentioned earlier, course through our bodies. As they are now, we were once and vice-versa. Like blood we are renewed - given heart by our good experiences as our old, tired, worn out bodies are crushed by our bad ones. Round and round we go, like peas in a boiling saucepan - rising and falling as we take what life has to offer and giving back what we've finished with. We eat and are eaten, beat and are beaten, in the great circle of life .

Sound is what relaxes us up and out - back into The Great Mystery as sight tenses us down and in, seeking understanding. Once we see the truth, we are released from the confusion of not knowing. We seek silence and stillness, when we want to know and noise or motion, when we don't. Like being at the bottom of the pool, after sinking like a lead weight, we grasp the trophy of enlightenment and rise gently to the surface again. Relaxing and contracting, we breathe the universe and it breathes us. We are one with the sea of life and it, with us. It is like God - within us, outside of us and part of all existence, when linked. To be at war with ourselves and all life, is to nothing and nobody but ourselves (our limits). To be at peace is to see our connection to 'the all' and to be a willing part of that 'all,' rather than being in conflict or competition with it (The only thing we ever truly fight is our own ignorance).

Jesus was lord of the dance because he gave into the rhythms of the universe. Those who fight reality, have two left feet. Their life is a staccato of mishaps because they continually put the brake on it but in order to see, they must do this. Once they have, they come out of their protective shell and move on.

To those who fear, what they crave, 'need,' more than anything is stability, to build up their shattered lives again (certainty, to calm the chaos and confusion they feel their world has been flooded with). To the bored, the opposite is needed - stimulation, not sedation. They are not sick of the world but sick of themselves and their limits. Life for them, for us has become a prison cell, not the haven of peace it once was. This is why we must move on. They have started the arrival process, which will ground them here as we are packed up and ready to go. Life holds no more mystery and confusion for us. It has all been laid bare for us and our roles have changed from students to teachers - masters of our own destiny, who no longer crave fame, fortune or position. We have had it and this perversely turns us into servants of something greater than ourselves. The future beckons - the past waves goodbye and we must walk on, where time leads us (We have all the time in the world to think and and all of space to act in. Those we try to help have none in their own minds, so lash out in anger and panic, fearing they will never achieve any of their aims, betraying themselves first and others trust in them, second). It is our negative beliefs that trap and trip us as our positive ones free us (Disbelief free us from any unwanted connections).

Some people consider The Earth a prison, others a school and some that it is a mental asylum, depending upon whether they feel it a hell, holding them here or one to release their potential. To those of us that stay behind in service to others, it is a hospital to mend bodies and minds in. It is your attitude that ensures the role it plays in your life from factory, to shop, farm, colony, science lab or theatre. To some it is a living organism of which they are just its cells in its body - to others it is an animal and we are parasites on its back. These visions or partial truths, allow many facets to exist in the minds and worlds of mankind. We are just witnesses to these different dreams of Man.

IS IT A DOG'S LIFE?

Here we are again. Shouted at for nothing. I'm a dog, doing what dogs do. It's in my nature but no, that's not good enough. So I'm sniffing another dog's backside - that's how we ask 'Who are you / how are you doing?' We smell the air and listen to the wind, to see what kind of day it is.

Spot, get out of that garden! Spot get off that other dog! For God's sake leave me in peace. If you want a wind up toy, get one - otherwise accept me for what I am.

What's with all this flower picking and sticking them in jars? Flowers are meant to be peed on, to let every other dog in the area know you've been there - that's why they are so tall. Good scent wafters and you leave them on the table until they die? What a waste and I don't think the plants are going to thank you for it either.

Get out of Mrs James garden! What is that all about? All this ownership crap. You humans are crazy! I'm a dog, a wild animal - we have no borders, no bars to exploration, unless it's defending something there and then. Even then, it's never anything more than on a temporary basis (Next year's den could be anywhere). The gypsies understand this way of life - you tamed people don't.

Keep off the flowers. Why, you're just going to pick them and stick them in jars, aren't you?

Don't pee on the roses. Don't jobby on the path. Stop humping Mrs Jackson's poodle - why she deserves it, the bitch. Stop chasing next door's cat, Willies pit bull etc. Can't

I do anything right according to you? Get off the settee! Don't roll in that! For God's sake, stop eating that, it's disgusting! Get your dirty feet off the furniture! Cleanliness is next to godliness - well dirt is next to dogginess. I feel sympathy for their kids but one day they'll grow up to be like their parents - all except Mr Kirkpatrick. Smelly Johnny, the loner, lives in a disgusting hovel, full of dogs and cats - lovely! He doesn't care about the dogs messing up the seats or cats scratching the covers to bits. Why couldn't I have ended up there? Paradise! Everyone curls up on the same dirty, stinky couches and eats out of the same bowls (a few squabbles) as well as drink out of the same water containers. Magic! Home from home. He doesn't think a dog is just for Christmas, not for life.

Instead here I am, locked in a house, lying on the floor but al least it is in front of the fire. Ah, man's greatest invention! Here I am wallowing in it. Spot, stop scrabbling your bed. But it's not comfortable? Why the fuss? Oh, you want to listen and look at your flashing box. Is that it?

Sitting there all night, except to get up and go pour boiling water of 'gag' liquid down their throats (whisky to you). Beer I can understand. Grass water - not too bad. I might even drink some myself, if given the opportunity. Then there's getting up every so often and going to the toilet. Flushing all that good stuff away, when you should be using it to mark out your territory, like drunks do on their way home from the pub. Tommy Kirkpatrick doesn't have to do this - everyone can smell where he is or has been. Disgusting old man! The frumpy ancient ladies say but us dogs say Eau De Cologne. Lovely - not like that horrible stuff you throw on yourself. Then you throw bits of cloth over your bodies, to further drown out any of your natural smells because you can't wash yourself clean , like we do (Cats are even better at it). Hey, stop licking your private parts! Why, do you want me to be as smelly as you? Humans!

Stop barking at the postie! Why, he's trying to get into the house and you're doing nothing about it. Someone has got to protect the place against invaders like him and obviously it isn't going to be you is it? Window cleaners, leaflet distributors, people collecting for charity - you'd let them all in but not me. I'm the man of the house and I've got to protect it.

There you sit, eating your corn flakes, pouring milk on them and scorn on me. And why do you spend half an hour each morning, staring at those sheets of black and white paper (coloured at the weekend)? More boiling water thrown down your throats and at night hot food tossed down there as well. Then you jump out of one square box (a house) and into another (a car), driving into work or to the shops. Meaningless rituals. Is that why you shout at us so much? Are you jealous of our perpetual childhood? Stop chewing the chair! Stop jumping up at the windows! (How else am I going to catch the flies - they're not going to come to me are they?).

Same outside. Stop chasing the birds, other dogs etc. Stop killing the voles and wasps. Stop this, stop that. My name is Spot, not stop.

Get your collar on. Get your lead on. Let me drag you here, pull you back there. Hold you back from chasing a cat. Still I got my revenge last Christmas - pulled the turkey off the work top. Got a beating for spoiling their celebrations but it made mine. Real food for once - not that dried muck or tinned stuff. Nice one!

Girls just want to have fun - well so do dogs.

CPSIA information can be obtained
at www.ICGtesting.com
Printed in the USA
BVHW050204090223
658191BV00031B/1018